THE

TWIN PARADOX

OF

SPECIAL RELATIVITY

IS

SOLVED

An Exploration of Special Relativity

by

John S. Mixson, PhD

The

Twin Paradox

of

Special Relativity

is

Solved

An Exploration of Special Relativity

By John S. Mixson, PhD

Zen saying - There are no paradoxes in Nature

The light postulate - The speed of light in empty space always has the same value. When a particular ray of light is measured from two reference frames in steady relative motion, both measurements produce the same speed c.

Copyright © 2014 by John S. Mixson, PhD

ISBN 978-1-4958-0016-0
ISBN 978-1-4958-0017-7 eBook

Printed in the United States of America

This book is a work of scientific investigation. It is based on previous work published in the open literature. The procedure is primarily interpretation, interpolation and extrapolation of the previous results, with some new calculations and explanations.

Published April 2014

INFINITY PUBLISHING
1094 New DeHaven Street, Suite 100
West Conshohocken, PA 19428-2713
Toll-free (877) BUY BOOK
Local Phone (610) 941-9999
Fax (610) 941-9959
Info@buybooksontheweb.com
www.buybooksontheweb.com

Table of Contents

Preface

Special relativity has been an interest of mine for a long time, since the subject began to appear in the newspapers, in the 1950's. But it was not until recently that I decided to make a special effort to understand it. Unfortunately, it seemed very difficult to gain a clear picture of relativity from the literature, both 'popular' and technical. Do clocks actually slow down, or is it only a perception of the observer? What is the relation between relativity and the rest of physics, particularly wave motions in acoustic media?

As a result of this situation I decided to do my own analysis of special relativity, picking up hints in various publications and doing my own math when necessary. My approach is to start with the very basics and to emphasize the connection of the physics of light and the math. The results of my analysis are presented here. I hope they will help other beginners who are having similar difficulties.

1. Introduction

Special Relativity continues to be a subject of considerable interest. Articles have appeared recently in popular magazines, science books, and college textbooks, References 1 to 4.. Searches in bookstores, libraries, and on the internet show that many more references are available. Lively discussions appear at websites where the general public can contribute.

Review of a considerable sample of publications shows that many emphasize the relativity phenomena known as length contraction, time dilation, and the twin paradox. The explanations presented for these phenomena seem incomplete. Attempts to dig deeper into the literature for better explanations were unsuccessful. Therefore a new, independent investigation was conducted, and the results are presented herein.

The idea of relativity was first presented by Galileo in 1632. He asserted that the behavior of objects in a ships cabin would be the same whether the ship was stationary or in steady motion, Ref.4, page 53. Sir Isaac Newton also advocated relativity, circa 1685, saying that the motions of bodies are the same among themselves whether the enclosing space is at rest or in uniform motion, Ref. 5, 6. The principle of relativity was then brought to bear on the theoretical problems where the physics of light merged with the physics of electro-magnetism.

Light was shown by Thomas Young in 1803 to exhibit properties, such as interference, that are characteristic of waves, Ref. 7. He also identified a material called the 'luminiferous ether' that was thought to be required to support the travel of the waves of light, in similarity with the air which supports the waves of sound. A series of experiments were conducted by Michelson and Morley to determine the effects of the motion of the earth through the ether, but no effects were found, Ref. 4,8. This result indicated that new ideas were needed to explain the properties of light. One attempt suggested that the experimental apparatus became shorter in the direction of motion due to the resistance forces applied by the ether. This idea would explain the experimental result and allow the ether to remain in the theory.

In the meantime in 1873 the previously separate equations for electricity and magnetism had been joined together into a single theory by James Clerk Maxwell, Refs. 9 and 10. This theory showed that light has the nature of electromagnetic waves, and calculated the speed of propagation of the waves. Maxwell's theory is considered a major historical accomplishment because it introduced the idea of the electromagnetic field and because it related all electrical and magnetic phenomena known at the time, Ref 10. Nevertheless deficiencies appeared after some additional research, namely the theory retained the idea of the ether (which could not be detected), and the equations were not invariant under a transform to other coordinates in steady relative motion (relativity was violated).

Around the time of 1900 it was considered of vital importance to correct the deficiencies of

Maxwells theory. Among the scientists working on this problem are included H. A. Lorentz of Holland, Ref. 11, H. Poincaré of France, Ref. 12, and Albert Einstein,Ref.-13. One gets the sense that all three were working toward the solution at similar rates. Ref. 12 indicates that Lorentz and Poincaré were corresponding regularly on the subject, and Einstein credits Lorentz for the equations of transformation. But Einstein is commonly given credit for relativity, perhaps because his ideas were closer to the physics of the problem Ref. 9, page 20. Einsteins theory was published in 1905 in the paper "On the Electrodynamics of Moving Bodies" Ref. 13. Motivations cited for the work include non-symmetries that appear when the electrodynamic equations of Maxwell are applied to moving bodies, and a desire to establish the validity of electrodynamic laws for the same coordinate systems for which the laws of mechanics are valid.

The non-symmetry is illustrated by envisioning a magnet and a nearby conducting wire, and describing the differences that occur when one, either the magnet or the wire, is considered to be in motion while the other is at rest. It is observed that the phenomena should depend only on the relative motion, and not on any idea of absolute rest. Einstein proposes to develop a new theory for moving bodies based on Maxwell's theory for stationary bodies and the following two postulates:

1. The Principle of Relativity: The equations of electrodynamics must remain the same when referenced to either of two coordinate systems in steady relative motion with respect to each other,

2. The Postulate of Constant Light Speed: The speed of light, c, is independent of the motion of the emitting body or of the receiving body.

The 1905 paper is divided into two parts, I. Kinematical Part, and II. Electrodynamical Part. The first part uses the Light Speed Postulate and careful consideration of the meaning of time and space to develop equations for the transformation of coordinates from one coordinate system to another in steady relative motion. These equations are generally referred to as the "Lorentz Transforms". The second part uses the Principle of Relativity to develop modifications to Maxwells equations, and explores the implications of those modifications, one of which is the equivalence of matter and energy expressed in the equation $E = m\,c^2$.

Ideas about slow clocks, shrinking objects and traveling twins appear to arise only from considerations related to the Lorentz Transforms, that is, the Kinematical Part. These ideas are the focus of the current study, so only Lorentz / kinematics will be considered here. Modifications of Maxwell's equations is much more complex and is not considered here.

The purpose here is to look at the basis of the theory in an effort to develop better understanding. Referring to the contents, the plan is to examine the details and the properties of the Lorentz transforms, with special attention to the connection between the equations and the physics of light, and then to address each of the special topics of interest.

2. The Lorentz Transforms

The purpose of this discussion is to highlight the relation between the terms of the Lorentz transforms and the properties of light, including the effects of the light postulate.

Special relativity envisions two Cartesian coordinate systems moving at constant speed with respect to each other. Neither is rotating, and they are oriented so that the x axis of each is directed along the direction of motion, as shown in figure 1. The x and x´ axes are coincident, but are drawn slightly apart to aid visualization.

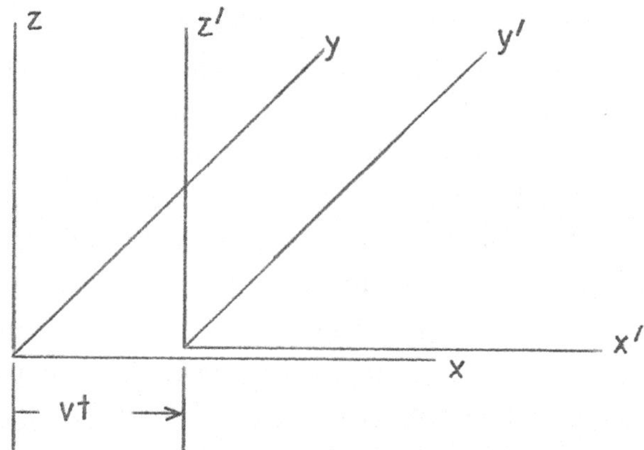

Figure 1.- Coordinates for relativity calculation

The space and time coordinates are chosen so that the origins coincide, thus $x = t = 0$ corresponds with $x´ = t´ = 0$. The coordinates are moving along their x, x´ axes with constant relative velocity v, therefore the origin of the x´ axis is located at x = v t, and the origin of the x axis is located at x´ = -v t´. Each system has clocks located at points of interest, and at rest in that system. The clocks are synchronized, using the exchange of light signals, so that when any clock indicates a time t all other clocks at rest in that coordinate system also indicate the time t.

In order to facilitate the discussion, the frame having the coordinates x,y,z,t is often referred to as the 'stationary' frame, and the frame having the x´,y´, z´, t´ coordinates is referred to as the 'moving' frame. These names are traditional but unfortunate because they suggest a preference. However, there is no place known to be at absolute rest, and both frames are inertial (not accelerating), so either one can be considered to be at rest. For each result obtained with the 'stationary' frame 'at rest', there is a symmetric result, equally valid, for the 'moving' frame 'at rest'.

The transform equations define the relation between the coordinates of the two systems when both are describing the same event. The event is considered to be instantaneous and therefore is described by specific space and time coordinates x, y, z, t, and x´, y´ z´, t´. Events that have been used as examples include the flash of a photographic bulb and the strike of a lightning bolt. The transform must satisfy the constraint that the speed of light c must be the same relative to each coordinate system.

It will be helpful to compare the relativistic Lorentz transforms with the non-relativistic Galilean transforms. Both apply to the coordinate pair shown in the figure.

Table 1. Transformation Equations

Galilean	Lorentz
$x' = x - vt$	$x' = m(x - vt)$
$t' = t$	$t' = m(t - vx/c^2)$
$y' = y$	$y' = y$
$z' = z$	$z' = z$
	$m = 1/\sqrt{(1 - v^2/c^2)}$

The Galilean transforms are familiar from everyday life. If x is the distance to our destination, then the distance still to go, x', is x minus our speed v times the time we have been traveling, t. If one vehicle is moving at speed u, and a second vehicle is moving at speed v, in the same direction, then the relative speed of the vehicles is u − v. These relations are modified at high speeds where the Lorentz equations apply.

Comparing the two transforms shows that the y and z equations are the same, the Lorentz transforms include the multiplying factor m in the x and t equations, and the Lorentz equation for time includes the additional term vx/c^2. These additional terms appear in the equations due to the requirement that the light postulate must be satisfied. The following analysis will study the relation between these terms and the properties of light.

In the limit, when the speed of motion v is very small compared with the speed of light c the Lorentz equations reduce to the Galilean.

Shift of the time origin

The meaning of the added term in the time transform can be described as follows. First consider the effect of the Galilean transform.

This part of the analysis takes the x, t system as the reference system. A series of events located at x = xo, shown in Fig 2a, forms the dotted line parallel to the t axis. The position of a point in the x´, t system is related to the x, t position by the equation

$$x´ = x - vt. \tag{1}$$

Therefore the dotted line in Fig 2a corresponds to the sloping line in Fig 2b given by the relation

$$x´ = xo - vt \tag{2}$$

Thus the effect of the motion is to add a negative slope to the line defined in the x - t system.

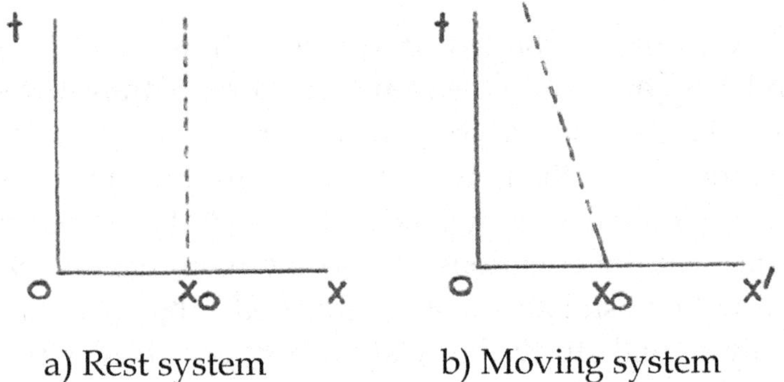

a) Rest system   b) Moving system

Fig. 2 A series of events

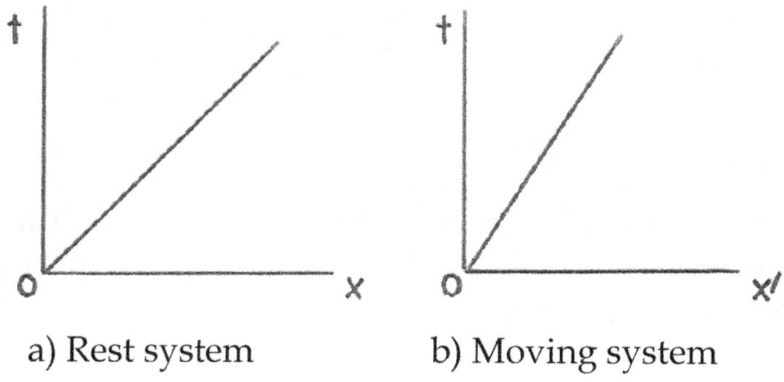

a) Rest system   b) Moving system

Fig. 3 Path of a light ray

The behavior of a light ray is shown in Fig 3. Fig 3a shows the path of a light ray in the x, t system, the system at rest. The equation of this line is

$$x = ct. \tag{3}$$

This relation describes the position of a wave front resulting from a light flash at $x = t = 0$. The path of the wave front relative to the x´, t system, the moving system, is shown in Fig 3b. Since the origin of the moving system is moving in the same direction as the light ray, the light ray doesn't travel as far relative to the x´, t coordinate origin, i.e. the motion adds a backward slope to the path of the light ray. The path of the light ray is shown by the solid line in Fig 3b, with the relation

$$x´ = ct - vt, \tag{4}$$

where the relation $x = ct$ has been inserted in eqn (1). So this shows that the light ray is not moving at the same speed in x´, t as in x, t.

But the light speed postulate requires the speed to be the same with respect to both coordinate systems. To enforce the constraint of constant light speed introduce the new time coordinate

$$t' = (c - v) t / c \qquad (5)$$

Solving eqn. (5) for t and inserting the result into eqn (4) shows that the constraint has been satisfied, since now

$$x' = c\, t'. \qquad (6)$$

Note that t′ is always less than t by the factor (1 - v / c). Eqn (5) can be written, using eqn (3), as

$$t' = t - vt / c, = t - v x / c^2, \qquad (7)$$

and so the transform (so far) can be written

$$x' = x - v t \qquad \text{a)}$$
$$\qquad (8)$$
$$t' = t - v x / c^2. \qquad \text{b)}$$

The significance of the second term in the time equation is as follows. The origin of the x′, t′ system is located at x′ = t′ = 0. Setting these values in equations (8) shows that the origin of the moving system is located, relative to the x, t system, at the position:

$$x,o = v\, t, \qquad t,o = v x / c^2. \qquad (9)$$

These values result also from a similar analysis of the complete Lorentz equations.

The value of x,o was established in Fig. 1. The value of t,o arose due to the coordinate change, equation(5), that was needed to satisfy the light speed postulate. Thus the origin of the time coordinate shifts forward as the light wave propagates, just as the origin of the space coordinate does.

Magnification factor, m

As shown in Table 1, the Lorentz transforms include the factor, m, termed the magnification factor, that does not appear in the non- relativistic Galilean transforms. The following analysis is intended to display the physics of light, as expressed in the Postulate of Constant Light Speed, that produce the factor m.

This analysis includes the following features, namely: the coordinate systems shown in Fig. 1, the assumption that the relations between coordinates are linear; and finally the analysis of a light pulse emitted from the common origins of the axes, reflected from a point at a fixed position in the moving system, and received back at the origin of the moving system, as shown in Figure 4.

Attention here is restricted to the x - t plane. As has been noted, the equations relating the coordinates x, t with the coordinates x', t' must be linear because of the homogeneous nature of space and time. Thus the equation relating t with x' and t' can be written:

$$A\,x' + B\,t' + C\,t + D = 0. \qquad (10)$$

Graphically this equation represents a plane surface, where t is a function of x' and t'. For a plane, determination of the values of the constants

A, B, C, and D requires that three points be known. The three points chosen are described as follows. A point is chosen at a fixed value of x´, at a distance L from the origin, where a mirror is placed. A light ray is emitted from the origin of x´, travels to the mirror, and is reflected back to the origin. The three points are 1) the departure of the ray from the origin, 2) the reflection of the ray from the mirror, and 3) the arrival of the ray back at the origin, as shown in Figure 4.

With the help of Figure 4 and the Postulate of Constant Light Speed, corresponding values of x´, t´, and t can be determined for each of these three points. It may help to visualize an old fashioned wooden ruler of length L at rest at the origin of x´ with the mirror attached to its end.

Point 1.- The origins of the coordinates are arbitrary, so they are set to zero. Therefore $x´ = t´ = t = 0$, and $D = 0$.

Point 2. - The Postulate of Constant Light Speed says that light travels at the speed c regardless of the motion of the source of the light and regardless of the motion of the observer of the light. Thus light travels at speed c with respect to the coordinates x´, t´, and the time taken to reach $x´ = L$ is $t´ = L / c$. However light also travels with speed c with respect to the x, t coordinates, and the ruler is also in motion at speed v. Thus the light must catch up with the ruler in motion in addition to traveling the length of the ruler. The time taken to

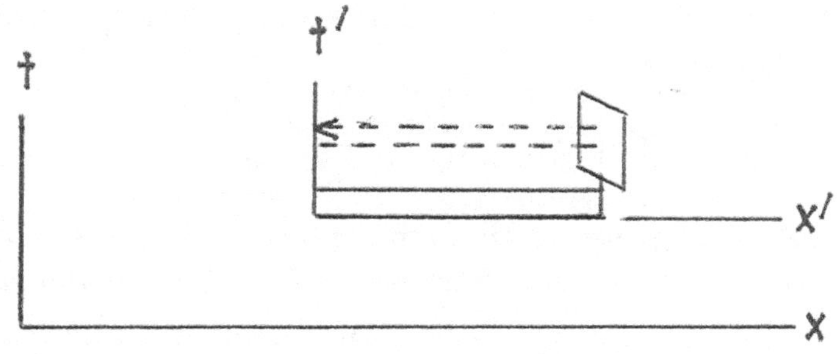

3) Return to origin of x´

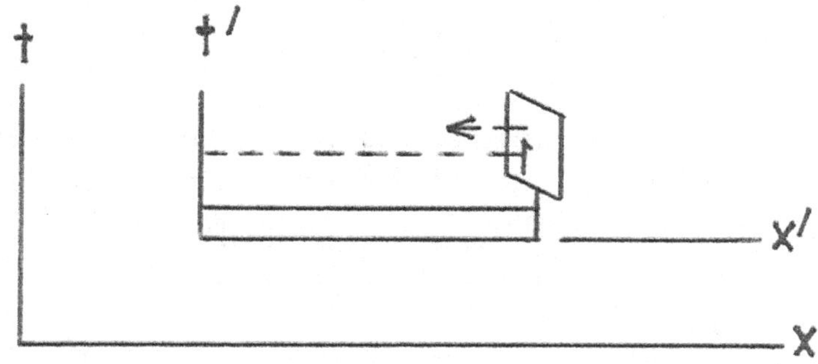

2) Reflection at x´ = L

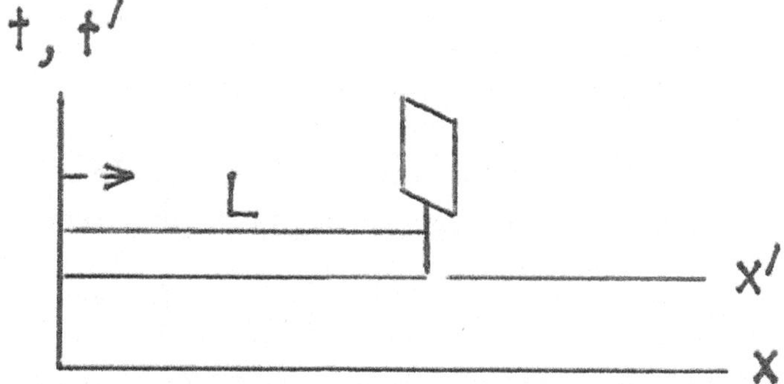

1) Light leaves the origin

Figure 4. Three points define t (x´, t´)

do this is $t = L / (c - v)$. The corresponding values are therefore

$$x', t', t = L, L / c, L / (c - v).$$

Point 3. - Applying the same reasoning as used for Point 2, and noting that the light and the ruler are traveling in opposite directions from the mirror to the origin, leads to the following corresponding values

$$x', t', t = 0, 2 L / c, L /(c - v) + L / (c + v).$$

Three simultaneous equations can be formed using Eqn 10 and two additional equations obtained by inserting the values for points 2, and 3 into Eqn 10. The coefficients of the constants A, B, and C are:

$$
\begin{array}{llll}
x' & t' & t & = 0 \\
L & L / c & L / (c - v) & = 0 \quad (11) \\
0 & 2L / c & 2Lc / (c^2 - v^2) & = 0
\end{array}
$$

These equations can be simplified by dividing the last two through by L. The solution is found to be

$$t = m^2 (t' + v x'/ c^2). \qquad (12)$$

Note that this equation gives t as a function of t' and x' whereas the Lorentz equations, Table 1, give x' and t' as functions of x and t. Elimination of x from the Lorentz equations gives the result corresponding to equation (12)

$$t = m (t' + v x'/c^2). \qquad (13)$$

The difference is the appearance of m^2 in (12) where m appears in equation 13.. Review of Ref 13 shows that m^2 appears there in the result of the analysis of one dimension, but when the y and z coordinates are accounted for the factor reduces to m.

Equation (12) shows that the magnification factor m and the origin shift, (the second term in the equation) both appear together from this single analysis method, and not separately from different physical or calculation processes.

The origin of the magnification factor can be seen with the help of Figure 4 and the following considerations. The two coordinate systems are moving with respect to each other, and both are "observing" the progress of the same light ray. But one observer is at rest with respect to the points of origin and reflection of the light ray, and the other is not. The Postulate of Constant Light Speed specifies that each sees the same speed of light, c. The magnification factor m then arises because the time of travel to the mirror and back is greater for the observer who sees the 'ruler' moving, so the light must catch up to the ruler in addition to traveling its length.

Space *x* time = distance ?

Embedded in the development of equation (12) is a remarkable new result, brought about by the light speed postulate. The time taken by the light when traveling between the points shown in fig. 4 is found from the relation

speed *x* time = distance.

The specific relations for fig. 4 are:

point 1 to 2:

$$c \, x \, t' = L \quad c \, x \, t = L + v \, t,$$

point 2 to 3

$$c \, x \, t' = L \quad c \, x \, t = L - v \, t.$$

Note the difference between x - multiply and x - coordinate. Calculation verifies that these relations lead to the time values given for the three points.

Comparing the times for the point of reflection, point 2, the time t is shown to be greater than the time t'. However the light ray has traveled farther with respect to x, namely a distance L + vt compared to just L with respect to x'. Therefore the greater time t = L/ (c - v) is consistent with the greater distance and together they describe the light speed c. On the return trip from mirror to origin of x', the distance traveled is L - vt, which is less, and the time of travel is L/(c + v), which is also less, and the speed is still c. In each case a longer distance corresponds to a larger time.

However, this is not the case for the total light travel from the origin to the mirror and back to the origin. The total distance traveled is the same for each coordinate system, namely 2L, found by adding the right hand sides of these equations. But the elapsed time is different for each coordinate system. At point 1 the clocks of both systems were set to zero. Therefore the values of t and t' at point 3 represent the time elapsed as the light ray traveled from the origin to the mirror and back to the origin of the x' coordinate. As previously given

$$t' = 2L/c, \text{ and } t = 2L/c. \, c^2/(c^2 - v^2). \quad (14)$$

Therefore the time elapsed in the x, t system is greater by a multiple of the factor m^2. The total trip time to the mirror and back is greater for observer x, t because the increased time outbound overpowers the decreased time inbound. This is the new result, namely that a light ray that travels the same distance at the same speed with respect to each of two moving coordinate systems does not necessarily take the same time. The principles of Fig 4 show how this result comes about.

Thus the ordinary, non relativistic, relation of time, distance, and speed does not hold in general for relativistic conditions. This result is in contrast with classical wave mechanics. Fig. 4 can be used to represent a wave problem for an acoustic or water surface wave. The time elapsed is then found to be the same for both coordinate systems.

<u>Summary</u>

The analysis associated with Figure 4 represents the fundamental mechanism for accounting for the light speed postulate in the Special Theory of Relativity. A similar construction is used in Ref 13 in a rigorous, calculus approach that accounts for a light ray moving with components in all three coordinate directions. The origin of the additional terms in the Lorentz transfoms is not apparent, however, without special detailed analysis.

The meaning of the terms in the Lorentz equations can now be summarized as follows. Referring to the x´ and t´ equations in Table 1, the first term on the right indicates the position of a point in the x, t system. The second term indicates the position in the x, t system of the origin of the x´, t´ coordinate system. The position of the origin is associated with the values of x and t, and the origin moves when they change. This connection between the origin and the values of x and t ensures that a light ray in one system transforms into a light ray in the other system.

The connection of the independent variables with the position of the moving origin also leads to the idea that the transform equations form a kinematic mechanism, such as a pantograph.

The magnification factor is now seen to account for the intricate relation of the reflecting light rays as observed in the two systems, as shown in Figure 4.

3. Dimensions, Clocks, and Rulers

The transform equations, Table 1, are dimensional equations. Therefore, the dimensions of the terms on both sides of the equations must match.

In the equation relating the space coordinates x and x′ the units of x and x′ must match; if x is measured in meters, then x′ must also be measured in the same meters. And the origin of the measurement must be the origin of the respective coordinate frame. If there are 'measuring rods' used to determine the values of x′, then those rods must have the same length as the rods used to measure x.

In the time equation, the units of t and t′ must match. If t is measured in seconds, then t′ must also be measured in the same seconds. And the origin of time must be the same so that when t = 0 also t′ = 0. And if there are clocks measuring t′ those clocks must count off time at the same rate as the clocks measuring t, and must be set so that t = 0 corresponds with t′ = 0. A clear description of this requirement is given in Ref. 5.

The light speed also has dimensional properties.. The requirement that light speed must be the same in both coordinate systems can be written mathematically as

$$\text{light speed, } c = x\,/\,t = x'\,/\,t'.$$

This relation states that the numerical value of c and the units in which it is expressed must be the same for both systems. If c = x / t is 186,000 miles per second then $x'\,/\,t'$ must give the same value. The relation can be satisfied if x' and t' are larger or smaller than x and t, but the proportions must be the same and the units of measure, miles and seconds, must be the same for x' and t' as for x and t.

In the derivation of the transform equations, the length of the 'ruler' holding the mirror is taken to be the same by both observers, and in general the analysis of Fig 4 is done with the idea that space and time measures are the same for both coordinate frames.

This discussion of dimensions might be unnecessary or self-evident when using the Galilean transforms because the results can be supported by intuition and experience. But in the case of the Lorentz transforms, the results are new and strange so there is little experience and intuition is unreliable. Careful consideration is therefore required.

4. Definition of Time

Special Relativity is concerned with the transformation of the Maxwell equations between the two coordinate frames shown in Figure 1. The definition of time is central to that transformation. The definition is based on the postulate that the speed of light is independent of the motion of the emitting body and on the existence of clocks located at rest in each frame. As previously described, the clocks of both frames are 1. started at zero together so $t = 0$ corresponds to $t' = 0$, and 2 are set to measure time at the same rate so the passage of one second on the clocks of the stationary frame (referred to as the K frame) corresponds to the passage of one second on the clocks of the moving frame (referred to as the K' frame).

The time t' is then defined as follows. Let a light ray start at the origin of the K frame in an arbitrary direction and reach the point x,y,z at time t. Accounting for the motion of the K' frame, coordinates x',y',z' can be determined that correspond to the point x,y,z. The time of the moving frame is then defined as the time on the clocks of K' when the light ray reaches the point x',y',z'.

Consider the following example. Let a light ray start at the origin of the stationary frame K(x,t), move along the x axis, and reach the location x at time t. During this time the origin of the moving frame advances the distance v t, leaving the distance x-v t for the light to travel along the x′ coordinate of the moving frame K′(x′, t′) to reach x. The point x′ = x − vt is then the point of the K′ frame that corresponds with the point x of the K frame. The light postulate says that the light travels at speed c in both frames, so x = c t, and x - v t = ct. Combining these equations leads to the time of the moving frame

$$t' = t - v\,x/c^2.$$

This equation corresponds to an approximation of the Lorentz transform for v/c<<1 which reduces the magnification factor to the value 1.

Paradigm shift In previous times coordinates and time were established by considerations outside the physics and were not changed in the analysis. Time was said to flow of its own accord. But in relativity the clocks and rulers are drawn into the action by the use of the light postulate, and their values determined by the light that they are measuring. This new perspective is different enough to constitute a 'paradigm shift'.

5. Time Network

The light postulate, the definition of time, and the mechanism illustrated in Figure 4 appear to be primarily a means for comparing the space-time properties of two relatively moving systems, and are not necessarily incompatible with the idea of a universal time that passes at the same rate at all places and for all observers regardless of their relative (uniform) motion.

The idea that the clocks of both K and K' are counting time in unison leads to the following interesting view. Imagine a third frame moving with respect to both K and K'. By the above reasoning, its clocks must match those of K and K'. By extension, the clocks of any number of inertial frames in relative motion must also match the clocks of K and K'. The view that arises is of infinite space, with innumerable inertial frames all in motion relative to each other having clocks that are marching in unison. No one frame is preferred over the others, and the clocks of any one frame are synchronous (in the sense of the exchange of light signals) only with other clocks of the same frame. However, this network of frames and clocks forms the foundation, or background, against which the motion of light rays is judged on equal terms for any two frames.

6. Transform Properties

<u>Transforms</u>

It will be convenient for the following discussion to state the transforms here.

Lorentz:

$$x' = m\,(\,x - v\,t\,),\ y' = y,\ z' = z,\ ct' = m\,(\,ct - v\,x\,/\,c).$$

For calculations starting with the x',y',z',t' coordinates it is convenient to solve now for the inverse transforms.

Inverse Lorentz:

$$x = m\,(\,x' + v\,t'\,),\ y = y',\ z = z',\ ct = m\,(\,ct' + v\,x'/\,c\,).$$

<u>A general light ray</u>

The transforms were derived so that any light ray in the stationary frame transforms into a light ray in the moving frame. This can be demonstrated as follows. A light ray in the stationary frame is described by the equation

$$x^2 + y^2 + z^2 = c^2\,t^2 \tag{15}$$

Substituting the inverse transforms leads to the result

$$x'^2 + y'^2 + z'^2 = c^2\,t'^2. \tag{16}$$

Thus a light ray originating at the origin of the stationary frame transforms into a light ray from the origin of the moving frame. It can also be shown that a light ray originating at any point in either system transforms into a light ray in the other system.

The speed of the ray is the same for both frames, but the direction and length of a ray are different, as shown in Figures 5 and 6. In these figures the calculation starts with given values of the x,y coordinates, shown by the 'x' symbols, and uses the Lorentz equations to obtain the x',y' values, shown by the 'o' symbol.

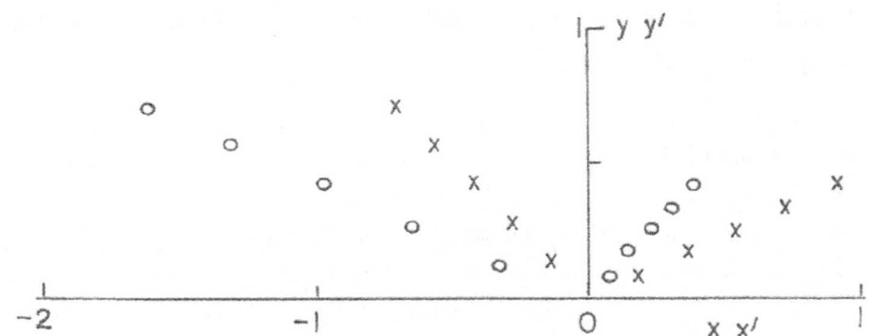

Fig 5 Transform of Light Rays. x - coordinates x, y

o - coordinates x ´, y´

Figure 5 shows that the the light rays, as located relative to the moving frame, are swept backwards along the direction of motion. The rays are shortened or lengthened depending on the forward or backward direction of the x, y ray. Thus the constraint of the light postulate introduces a non-symmetric distortion of the light rays.

A more complete picture of the relation is shown in Fig. 6, where 13 rays of unit length are evenly spaced in the x,y frame. This figure shows that the evenly spaced array of the x,y frame transforms into an unsymmetrical array of varying length rays in the x′,y′ frame.

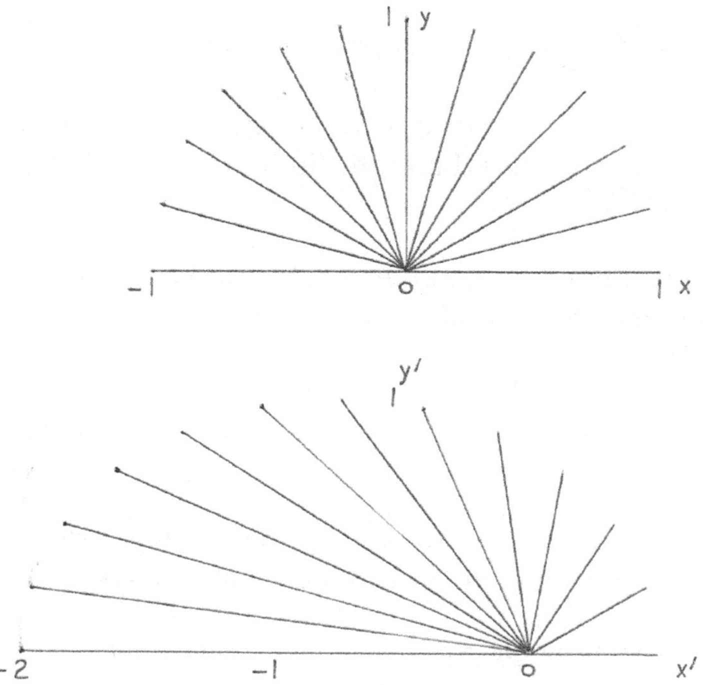

Figure 6.- An array of light rays transformed.

Transform Magnitudes

The magnification factor m has a value that is equal to or greater than 1, and approaches infinity as v approaches c. However, transformed light rays can be either larger or smaller than the original, depending on which system is taken as reference, and whether the light ray is traveling in the forward or the negative direction. The magnitude depends on both m and the other terms in the transform equations.

For a forward moving light ray, $x = c\,t$. Substitution into the transform equations leads to

$$x' = x\,\sqrt{(1 - v/c)/(1 + v/c)},$$

$$t' = t\,\sqrt{(1 - v/c)/(1 + v/c)} \qquad (17)$$

Thus, for non-zero positive v, x´ and t´ are always smaller than x and t. In the limit of $v = c$ both x´ and t´ become zero.

For a ray moving in the negative direction, $x = -c\,t$, and the corresponding equations are

$$x' = x\,\sqrt{(1 + v/c)/(1 - v/c)},$$

$$t' = t\,\sqrt{(1 + v/c)/(1 - v/c)} \qquad (18)$$

For this ray the coordinates of the moving system, x´, t´ are greater than those of the stationary system, x, t.

<u>The Moving Origin</u>

The Lorentz transform equations contain a unique and meaningful feature, namely a single symbol is used to represent two different physical concepts. The symbol x is used to represent the position of the independent variable, and to represent the position of the origin of the moving coordinates. Similarly, t is used to represent the independent time coordinate and the time location of the moving coordinates. (Taking x, t as the 'rest' system and x´, t´ as the moving system.) The coordinates of the origin of the moving system are given by:

$$x, \text{origin} = v\,t, \text{ and } ct, \text{origin} = v\,x/c \qquad (19)$$

It can be seen that the selection of specific values for the independent variables x and t also establishes an exact position of the origin of the moving frame. To emphasize, the position of the moving origin and the position of the event identified by the independent variables x and t cannot be specified independently; the positions are rigidly tied together by the transforms.

This relation between the origin and the independent variable can be helpful in understanding and picturing the mechanism of the transforms. An example for the case of a light ray originating at the common origin of the two frames is illustrated in Fig. 7.

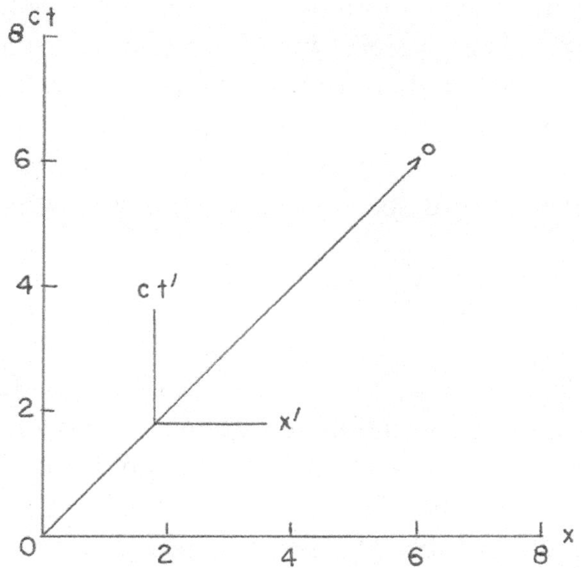

Figure 7. - Transform mechanism for v/c = 0.3
x, o =c t, o = 1.8, x´ = ct´ = 4.4

For this case the ray has progressed to x = ct = 6, indicated by the arrow. The corresponding position of the moving axes is x = ct = 1.8. The value of x´ = ct´ = 4.4, when added to the 'prime' axes, is shown by the small circle. This figure shows that the origin of the moving system follows the light ray along the line x = ct, and that the values of x´, ct´ fall on that line also, indicating that x´ = ct´, as the equation derivations were intended to accomplish.

Figure 7 shows that the moving, 'prime', axes have advanced a significant distance, almost a third of the way to the wave front, the arrow. It can be expected for this movement to have a significant effect on the transformed values, x´, and t´. In the case shown in Figure 7 the effect is a reduction of the values. On the other hand, the magnification factor m is greater than one, and can reach very large values for values of v/c approaching one (1). It is of interest therefore to compare the effects of the axis movement with the effects of the magnification m. For the case x = c t these two factors are expressed separately in the equations:

$$x´ = m\,x\,(\,1 - v/\,c\,)\,x\,x$$

$$(\,20\,)$$

$$ct´ = m\,x\,(\,1 - v/c\,)\,x\,ct$$

The term in parentheses represents the movement of the 'prime' axes, and m is the magnification factor given previously. The magnitudes of these two terms are shown in Table 2 for some values of v/ c.

Table 2 - Factor Magnitudes. x = ct = 6:

v/ c	0.3	0.8	0.999	0.9999
m	1.05	1.67	22.36	70.71
1 - v/ c	0.7	0.2	0.001	0.0001
x´, ct´	4.4	2.0	0.134	0.042

The term m tends to increase x´, ct´ while the axis shift term 1 -v/ c tends to decrease them. The table shows that the axis shift dominates for all v/c, even when m becomes very large. This result confirms the result of equation 17 that the transformed coordinates tend to zero as v/ c approaches 1.

Thus a large value of the magnification m does not necessarily mean that the transformed values will be also large.

Additional values of the coordinates x. ct are compared with the transformed x´, ct´ in Figure 8.

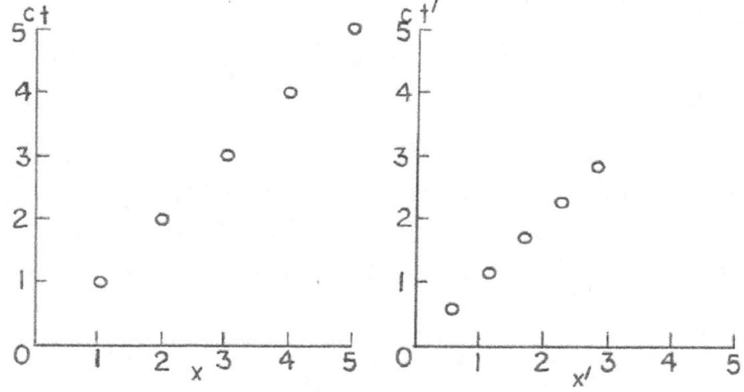

Figure 8.- Comparison of x, ct, at rest vs. x´, ct´ moving. v/ c = 0.5

In Fig 8 separate graphs are shown for x, ct and for x´, ct´. The points in the x´, ct´ graph are the transforms of the points in the x, ct graph; the point closest to the origin of x´, ct´ is the transform of the point closest to the origin of x, ct, and so on. The values of x´, ct´ are smaller than those of x, ct. The reduction is primarily due to the position of the origin of x´, ct´ as indicated in Table 2. The movement of the time origin to match the space origin in order to satisfy the light postulate is essential to these results..

Symmetry and Reciprocity

There is a kind of symmetry between the 'moving' frame and the 'stationary' frame. The absence of any substance, such as the ether, that transmits light rays leads to the idea that there is no one special frame of reference. (For mechanical waves, such as sound waves, water waves, or seismic waves, the frame at rest with the medium, air, water, or earth, is the special frame.) The idea of absolute rest is excluded, and each frame is equivalent for reference purposes.

Reference 13 considers an object of spherical shape as described by the coordinates of the moving frame. When those coordinates are transformed into the coordinates of the 'stationary' frame the object appears shortened in the x direction as described by the stationary coordinates. The object is said to be shortened 'when viewed from the stationary frame'. The symmetry is indicated by the observation that a sphere at rest in the stationary frame will also appear shortened 'when viewed from the moving frame'.

Figure 7 pictures a light ray moving in the +x direction, so x = ct and the reference frame is the stationary frame. The symmetrical figure when the moving frame is the reference frame has a light ray moving along the - x' direction, so x' = - ct'. Review of the inverse transforms shows that in this case the x,ct coordinates are smaller than the x', ct' coordinates. The symmetry is expressed as 'when a light ray moves in the direction of the non-reference frame, the non-reference coordinates are smaller than the reference coordinates.'

This kind of symmetry appears in other examples to be described in following chapters..

However, the coordinates of the two frames are reciprocal in the sense that any point x,y,z,t transformed into x',y',z',t' using the Lorentz equations transforms back to the original point using the Inverse Lorentz equations. This can be verified by combining the transform equations, and by the example of figure 8. Thus in Fig 8 each point in the x', ct' graph transforms back into its original, larger, x, ct coordinate values using the inverse transformations. The reason for this can be seen by reviewing figure 7. No matter which system is taken to be at rest, the x', ct' system is closer than the x, ct system to the event, which for this figure is the arrival of the light ray at the position x, ct = 6, therefore its coordinates describing the event are smaller.

Three dimensional radiation

The previous analysis has been concerned with the transformation of a single light ray. But in general a single light source will radiate in several directions at once, perhaps even in a spherically symmetric shape. The Lorentz transforms have been derived considering rays of light moving in various directions, but it is not clear what relation the rays have with each other. The analysis of Ref. 13 indicates that a spherically radiating light wave originating from x, y, z, t = 0 transforms into a spherical wave originating from the moving origin $x´$, $y´$, $z´$, $t´$ = 0. The analysis given there is quite brief and does not show how the mechanisms of the Lorentz transforms bring about this result. Therefore, it is of interest to examine the transformation of a wave that spreads spherically. The aim is to obtain a clearer understanding of the mechanisms of the transforms, i.e. the geometric relations, to show how the light waves and the coordinate systems fit together.

This analysis is based on conditions previously described. A short-duration flash of light occurs at the origin of the x, y, z, t coordinates. It is uniform spherically, so a spherical shell of light (or the wave front) spreads with speed c. Following Huygens principle the shell or wave front retains its spherical shape as it expands according to the relation

$$x^2 + y^2 + z^2 = c^2 t^2.$$

A second coordinate system, $x´$, $y´$, $z´$ is located with it's respective axes parallel to x, y, z, $x´$ being coincident with x, and the whole system moving with speed v along the x axis, Fig. 1. The origins of the two systems coincide when the light flash is emitted.

The Lorentz Transform determines the primed coordinates when specific values of x, y, z, t are given. The following exercise will determine the primed values for the specific sphere when ct = 1. Points in the plane z = 0 transform into points in the plane z′ = 0, and the light sphere and its transform are symmetric for y = positive and y = negative, so it will be sufficient to consider only the arc z = 0, x from -1 to +1, and y greater than 0. For this example v = 0.6 c.

The original wave front semi-circular arc, given by

$$x^2 + y^2 = ct^2 = 1, \tag{21}$$

is shown in Fig. 9, along with the transform of this curve into the coordinates x′, y′, ct′. Points on the circular arc are transformed using the Lorentz equations with y′ = y, and are shown as the elongated curve, Fig 9A, and time ct′ curve, Fig 9B.

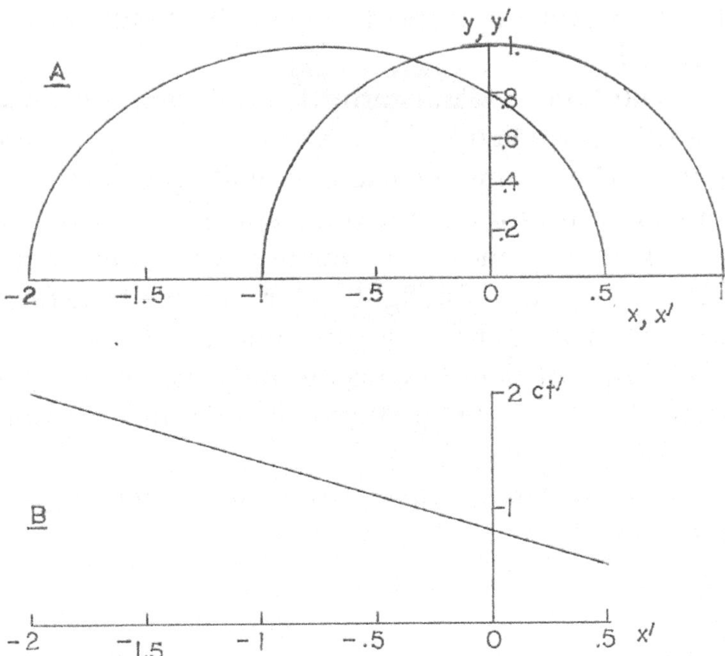

Fig. 9 - Transform of Unit Circle, ct=1, v/c=0.6

The shape of the transformed curve can be obtained analytically as follows. Substitute $y = y'$, and $x = x'/ m + v t$ (from the transforms) into equation 21 and rearrange to obtain:

$$\frac{(x' + m v t)^2}{(m c t)^2} + \frac{y'^2}{(c t)^2} = 1 \qquad (22)$$

This is the equation of an ellipse with center at (- m v t, 0) and semi axes of (m c t, c t). Calculations show that this curve describes the elongated curve of fig 9A. This result shows that a spherical wave front in x, y at a fixed value of ct does not transform by the Lorentz Transforms directly into a spherical wave in x', y', but rather into an ellipse.

For the value of m = 1 the above equation describes a circle with its center at (x, y) = (-v t, 0). This is the equation to be expected for a circular wave in a mechanical medium when 'viewed' from a moving coordinate system. An example that is easily visualized is the circular wave generated by a stone dropped into still water.

As an aside, the light wave can be assumed to be spreading spherically in the coordinates x′, y′, z′. Analysis parallel to that described above shows that this light wave also transforms into an ellipse with foci on the x axis, i.e. the curve is elongated.

Fig 9B shows the values of c t′ corresponding to the elliptical curve. This shows that the single value of ct = 1 transforms into a range of values of c t′. Thus there is no one-to-one correspondence between ct and ct′. The reason for this is related to the position of the moving origin, to be discussed presently.

The equation for ct′ can be obtained by eliminating x from the transform equations and rearranging terms to get

$$c\,t′ = c\,t\,/\,m - v\,x′\,/\,c \qquad\qquad (23)$$

Substituting ct = 1 and v / c = 0.6 shows that this equation describes the curve shown in fig 9B. Thus as ct increases, starting from zero, ct′ forms a family of parallel curves as shown in Fig.10. The values of x′ are limited by the size of the expanding sphere in x, y. The limits are shown by the dashed lines at + and - 45 degrees.

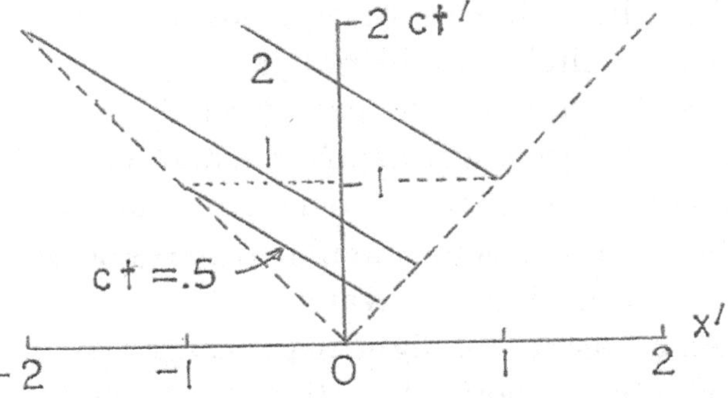

Fig 10 - Family of Curves at ct = constant

The shape of the curve of x′ vs y′ for a constant value of ct′ can be determined with fig.10 as a guide. Consider the value ct′ = 1. This value is intersected by sloped curves for ct = 0.5 to ct = 2 for all values of x′ from -1 to +1. So the procedure for finding x′ vs. y′ is as follows.

1. Find pairs of values of ct and x′ for a constant value of ct′, either by reading from fig 10 or by calculation from the equation

$$ct = m(ct′ + v\, x′ / c) \qquad\qquad (24)$$

2. Find the value of x corresponding to the pair ct, x′ using the equation

$$x = x′ / m + v\, ct / c. \qquad\qquad (25)$$

3. Use the values of x and ct thus found to find y′ from the equation

$$y′^2 = y^2 = ct^2 - x^2. \qquad\qquad (26)$$

The pairs of values, y′ from this equation together with x′ from step 1, are the required coordinates of the curve x′ vs y′ at constant ct′. This procedure has been carried out and the result is that the shape of x´ vs y´ for constant ct´ is a circle.

As observed from fig 10, values of ct from 0.5 to 2 are required to form the single circle in x´, y´ for the constant value ct´ = 1. There does not appear to be any guiding reason for the choice of a particular value of ct´, that is, following the transformation equations from the expanding sphere in the rest system through to fig 10 does not lead to any specific value of ct´ that corresponds to a particular value of ct for the whole sphere. The idea that an expanding sphere in the rest system corresponds to an expanding sphere in the moving system expresses only a formal relation, not a kinematical relation. Kinematical relations do exist for individual rays, however, as described below.

Referring again to fig 10, for the constant value ct´ = 1, the curves of constant ct intersect this line first (ct = 0.5) at x´ = -1. As ct increases the intersection moves forward, to increasing x´. The physical picture this suggests is of a spherical locus of points that is illuminated starting at the rear, with the illumination progressing forward in the shape of a circle in the y´, z´ plane, until the circle of illumination goes to zero at the forward end of the sphere, x´ = 1. That is, the sphere in the prime coordinates does not appear all at once, but only a circle at a time, as ct increases.

Moving Origin.- Further understanding of the transformed circle shown in Fig 9 can be obtained from the position of the origin of the 'moving' or primed coordinate system.

The origin of the moving system is exactly known. The x position is given in the problem statement, namely the moving coordinate system moves in the direction of positive x with speed v, and the origins of the two coordinate systems coincide when t = t´ = 0. The time position of the moving origin is determined by the time shift required to satisfy the requirement of the Postulate of Constant Light Speed, as previously discussed. The position of the origin has been given by:

x, o = ct v /c, and c t, o = v x /c.

These equations can be used to analyze the transforms shown in Fig 9. Consider first the ray traveling along the positive x axis with position x = ct. This ray is shown in fig 11, the letter A indicates the position of the ray at ct = 1. Using the relation x = ct and the values v /c = 0.6 and ct = 1 appropriate for point A, leads to the origin position x, o = c t, o = 0.6. An x´, ct´ coordinate system is shown at this position. Relative to this coordinate system the point A is located at x´, ct´ = 0.4, 0.4. Applying the magnification factor m = 1.25 that goes with v / c = 0.6 gives the final transformed values x´, ct´ = 0.5, 0.5. These values are shown in fig 9 at the right hand end of the ellipse.

Next consider the ray moving along the negative x axis, also shown in fig 11. The letter B indicates the position of this ray at ct = 1. Using the relation x = - ct and the other values given results in the origin position x, o = 0.6 ct, o = - 0.6. An x´, ct´ coordinate system is also shown at this position. Relative to this coordinate system the point B is located at x´, ct´ = - 1.6, 1.6. Applying the factor m = 1.25 gives the final transformed values x´, ct´ = -2, 2.

These values are shown in fig 9 at the left hand end of the ellipse.

These results show that the movement of the moving coordinates in both space and time, as required by the problem statement, provide a major part of the transformed magnitudes, for speed v / c = 0.6. The position of the moving axes is shown to be different for rays moving in two different directions. Since the position of the moving axes is forced by the kinematics of the problem statement, it follows that only one ray can satisfy the Light Postulate when several rays radiate at once. The solution of a transform when several rays are present thus requires that each ray be treated separately, with its own origin position. From the results presented above for the positive- and negative-going rays it is apparent that the Lorentz Transforms takes this feature into account.

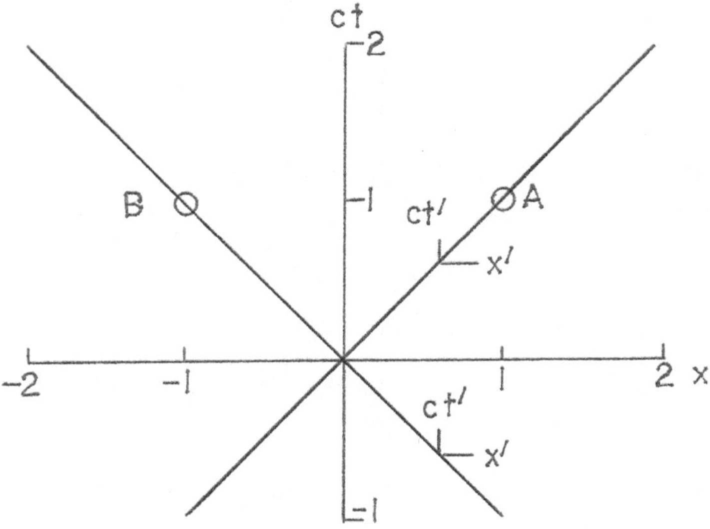

Fig 11. Origin Positions

7. Length Contraction

The Ether

The idea that solid objects become shorter when in motion arose following experiments that attempted to demonstrate the presence of the ether, Ref. 8, also 4, 18.

The experiments, ref 8, sought to determine the motion of the Earth through the ether. The apparatus featured a massive stone block upon which were arranged a light source, beam splitters, mirrors, and a detector for measuring the shifts of interference fringes due to the motion. Extensive tests were performed but no shifts of the fringes occurred. In an attempt to explain this 'null' result, while retaining the ether, it was proposed that the apparatus moving through the ether experienced resistance that caused the apparatus to shorten just enough to cancel the effect of the motion. This effect is known as the 'Lorentz-Fitzgerald' contraction. The ether was viewed as absolutely stationary so there was no consideration of a stationary Earth with a moving ether.

Relativity

Special relativity does not include the ether or any concept of an absolutely stationary space. The two postulates of 'The Relativity Principle' and 'The Constant Light Speed' are said to be sufficient for a consistent theory. Application of the postulates leads to the Lorentz transforms, which have been used to study contractions. The contractions found, however do not imply the shortening of solid bodies, as the ether theory does, but rather a variation of coordinate between the two frames.

For example, the transforms were used to describe a moving sphere, p. 33. The results indicated that the sphere 'appeared shortened when viewed from the stationary frame'. In the symmetrical case a sphere at rest in the stationary frame 'appears shortened when viewed from the moving frame'. This result is significantly different from the ether analysis.

There have a number of studies of contraction as depicted by Special Relativity. The results of these studies indicate that solid bodies do not contract due to motion. Any apparent change of length calculated is due to the method of measurement, Ref. 4, or to a particular viewpoint, Ref 18,20.

Shortening by Transformation

The use of the Lorentz transforms to compare lengths is illustrated in Figures 12 and 13. This analysis envisions a segment of the x axis for several values of the time ct, as shown in figure 12. For three values of ct a pair of points is shown, one at $x = 0$ and one at $x = 2$. These pairs of points represent a segment of the x axis of length 2 as time moves forward. The objective is to find the distance between the pairs when transformed to the coordinates x', ct'. The transformed points are shown in fig 13. Each point of a given pair transforms to a different value of time, ct', as well as to a different coordinate of space, x'. In Figure 13 each pair of points at the same value of ct is joined by a line, and the points having the same value of x are also joined by a line.

Figure 13 shows that two points at the same value of ct are separated by a greater value of x' than of x. Thus the segment of length x = 2 could be said to have expanded with the motion. However comparison can also be made between the lines of constant x. The separation between these two lines as measured at the same value of ct' is shown to be less than the x separation of 2.

Using the notation x1,ct1 to refer to the points at x = 0, and x2, ct2 to refer to the points at x =2, the transform equations can be written:

$$x1' = m(x1 - v\,ct1/\,c\,), \quad ct1' = m\,(\,ct1 - v\,x1/c\,)$$

$$(27)$$

$$x2' = m(x2 - vct2/c\,), \quad ct2' = m\,(\,ct2 - v\,x2/c\,).$$

Case 1: if ct1 = ct2 then (x2'-x1') = m(x2 - x1) and the transformed interval is larger. But the transformed time values, given by ct2' - ct1' = -m v/ c (x2 - x1), are not equal.

Case 2: if ct1' = ct2' then (x2' - x1') = (x2 - x1)/ m, and the transformed interval is smaller. This case seems preferred in the literature, e.g. Ref 19.

So the transformed spatial interval can be larger or smaller, depending on the treatment of the time interval. The symmetry between the two frames is shown by the result that the interval having both ends at the same time transforms to a larger interval.

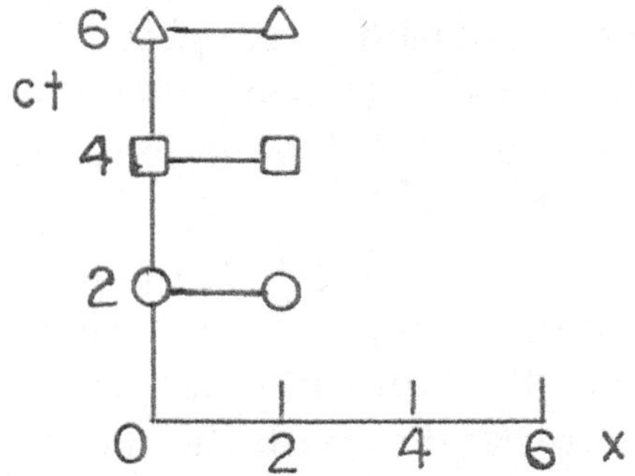

Figure 12 Pairs of points for contraction study.

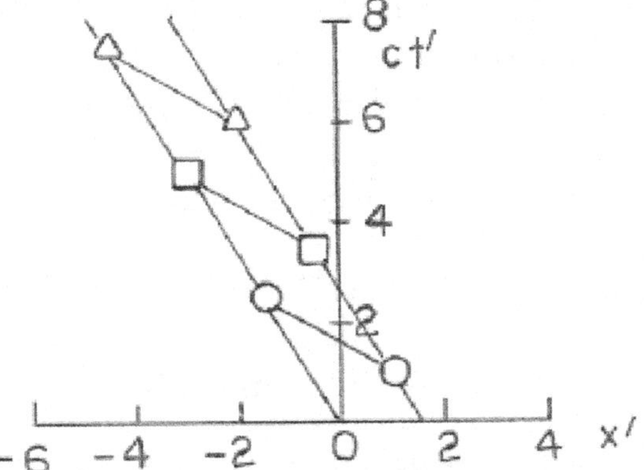

Figure 13 Transformed coordinates. v/c = 0.6

8. Time dilation - trajectories

Time dilation and 'moving clocks run slow' are two terms that refer to the result that transformed values of time ct´ can be less than the reference ct. There are several ways to analyze the situation.

<u>Conventional</u>

One way to calculate 'slow clocks' is as follows. Consider the clock located at the origin of the moving frame, $x´ = 0$, and at rest in that frame. With respect to the stationary frame, x, ct that clock moves following the relation $x = v\,t$, as indicated in Fig 1. If the point described by the independent variables x, t also follows the relation $x = v\,t$ then that point coincides with the clock. The value $x´ = 0$ can be substituted into the spatial transform $x´ = m\,(\,x - v\,t\,)$ to obtain the result $x = vt$, which can then be substituted into the time transform $ct´ = m\,(\,ct - v\,x/\,c\,)$ to obtain the result

$$ct´ = ct \sqrt{(\,1 - v^2/\,c^2\,)}. \qquad (\,28\,)$$

Thus the transformed time ct´ is less than the reference time ct. This analysis is presented in Ref 13, where it is noted that the time of the moving clock is slow as 'viewed in the stationary system'.

An example calculation is shown in Fig. 14. In this case the values of x, ct are considered as a time sequence, with x starting at 0 and increasing with time. (The value of x´, ct´ for a given x, ct does not depend on previous values of x, ct that may have been transformed.) The idea of a sequence is used to aid visualization of advancing time on the clock.

Therefore the times ct´ are considered as increasing readings on the clock at rest at x´ = 0. Such a sequence of values of x, ct is referred to as a 'trajectory'.

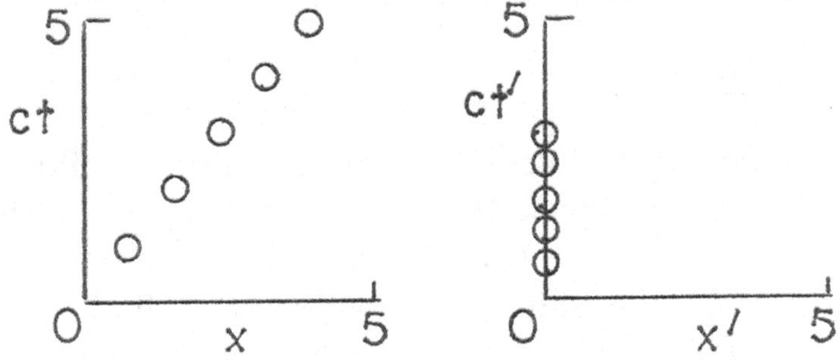

Figure 14. Sample of slow clock calculation.
v/c = 0.8 x = 0.8ct

Aside. The term time dilation suggests the idea that the 'seconds' of the moving clock have expanded so that their fewer number can match the time interval marked off by the more numerous 'seconds' of the stationary frame. In addition, the terms 'time dilation' and 'slow clocks' may suggest that something has happened to the moving clocks to cause them to operate incorrectly. As discussed in the section 'Dimensions' these ideas are not accepted in the present analysis. The observed features, such as 'slow clocks', and 'length contraction' are considered to result from the definition of 'time' and the properties of light as expressed in the Light Postulate and in the transform equations. The terms 'time dilation' and 'slow clocks' may be considered as shorthand for the relevant features of the transforms, which may not be simple to describe otherwise.

<u>Intervals</u>

The equations (27) used for the length analysis can also be used for time transformation. Let x1, ct1 refer to a point at a non-zero value of x and x2, ct2 refer to a point at a larger value of time.

Case 1: If x1 = x2 then the right hand pair of equations indicate that ct2′ – ct1′ = m (ct2 – ct1), and the transformed time interval is larger than the reference interval. But the transformed space interval (x2′ – x1′) is not zero as shown by the left hand pair of equations.

Case 2. If x1′ = x2′ the left hand pair of equations can be combined with the right hand pair to show that (ct2′ – ct1′) = (ct2 – ct1) / m, and the transformed time interval is smaller.

The symmetry between the two frames is shown here, as for length transformation, by comparing the two cases. The time interval where both ends of the interval are at the same value of x (either frame can be set this way) transforms to a larger interval. (In case 2 with x1′ = x2′ the transform is from the prime frame to the not-prime frame and the transformed interval is (ct2 – ct1) = m (ct2′ – ct1′).)

<u>Other trajectories</u>

The conventional analysis determines the transformed time associated with the trajectory x = v t. This trajectory has the particular property that each point on the trajectory transforms to the same value of x′, namely x′ = 0. The result, equation (28), is then said to represent the time on the clock at the origin of the moving frame.

Transformed time may also be calculated for other trajectories, however the individual points on these trajectories may not all transform to the same value of x′, i.e. each point may transform to a different clock. But, it may be recalled that the clocks of a given frame are synchronized together so that when one clock reads t = 10, for example, all the clocks of that frame also read t = 10. So, when the time of the moving frame is calculated, that time applies to all the clocks of the moving frame. Therefore, when reviewing the transformed time for various trajectories that time can be visualized as appearing on any single clock of interest, such as the one at x′ = 0.

For example, the transformation of time has also been obtained using the trajectory x = ct, Fig 7 and 8. The amount of slowing is given in equation 17, and is different from the values given in equation 28. The related discussion presents the role of the moving origin in causing the slowing. In addition, fig 4 and the related discussion explains the mechanism of the light postulate that leads to the transforms and the resulting time variations.

The trajectory that results in the relation

$$ct′ = ct.$$

is found as follows. Setting ct´ = ct in the transformation equation, ct´= m (ct - v x/ c), leads to the trajectory equation

$$x = ct (m - 1)/ (m v/ c). \qquad (29)$$

Thus, for each value of v/ c there is a trajectory along which ct´ = ct. Solving this equation for ct and substituting the result into the space transformation x´ = m (x - v t) leads to the result

$$x´ = - x. \qquad (30)$$

These equations can be applied to the transform of the unit circle as shown in figure 9. For v/c = 0.6 and ct =1, as used in that figure, equation 29 gives a value of x = 1/3, indicating that ct´= ct at x´ = -1/3. The line in Fig. 9B shows values of ct´. A horizontal line at ct´ = 1 (where ct´ = ct) intersects the ct´ line at about x´ = -1/3, in agreement with the above equations. This figure also shows clearly that the time of the moving frame, ct´, can be less than, equal to, or greater than the time of the stationary frame, ct.

A further example is shown in fig 15. In this case the trajectory is x = 0.5 ct compared to the speed of the moving axes v/ c = 0.8. The figure shows for each value of ct that the time of the moving system ct´ is equal to the time of the stationary system ct, and x´ is the negative of x.

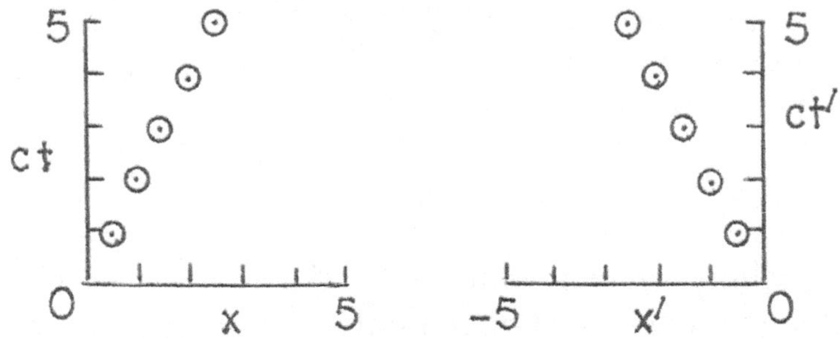

Figure 15. Equal Time transform. v/c =0.8, x = 0.5ct

From these examples it may be concluded that there is no one statement that describes the transformed time, but that it may be smaller, equal to, or larger than the original time.

9. The Twin Paradox

<u>Paradox:</u> A statement that is seemingly contradictory or opposed to common sense and yet is perhaps true. To resolve a paradox, find and correct the error in the reasoning.

<u>The Paradox</u>

The twin paradox as currently presented envisions two people who are twins. One twin leaves on a long space journey while the other remains at home. When the traveling twin returns home to earth it is calculated that he is younger than his stay- at- home twin. The paradox is the conflict between this calculated age difference and the expectation that the twins should remain at the same age.

The calculations take the stay- at- home twin to be at rest. Attempts to resolve the paradox by taking the traveling twin to be at rest are rejected, in the literature, by asserting that the traveling twin experiences forces and accelerations associated with the change of direction from the outbound motion to the homebound. Thus, the twins journey is not allowed the symmetry that otherwise exists in Special Relativity due to the equality of the frames that are both inertial, i.e. not accelerating.

The Resolution

The plan. The origin of these ideas about twins is apparently in Ref. 13, Part I, section 4. That analysis discusses clocks, which later writers represented as twins. The plan here is to follow the steps outlined there.

The View. The analysis starts by noting that a moving clock, when observed from the stationary coordinate system, records less time than an identical stationary clock. This result assumes that the moving clock follows a straight line, as indicated in Figure 1. The calculation starts by setting $x' = 0$ in the Lorentz transforms, indicating that the result applies to the clock at the origin of the moving frame. Applying this value in the space transform leads to the relation $x = v\,t$, which combines with the time transform to produce the 'slow clock' formula

$$t' = t\,\sqrt{(1 - v^2/c^2)}. \qquad\qquad (31)$$

This result is described as showing that the time of the moving frame t' is slow 'as viewed from the stationary frame'. The process of 'viewing from the stationary frame' consists of, first, description of a configuration of events as given by the coordinates x',y',z',t', then transformation of these coordinates into the coordinates x,y,z,t of the stationary frame, and finally the 'view' consists of the description of the configuration of the coordinates x,y,z,t. Viewing does not change anything about the moving frame, a moving observer experiences the same conditions whether viewing is taking place or not.

Muons. The principle of the 'slow clock' is supported by an experiment involving μ-mesons, or muons, Ref 18. Muons are charged particles produced high in the earths atmosphere by cosmic rays. They speed toward sea level, and some of them spontaneously disintegrate, or decay, as they go. They don't all decay at the same time, however. During a given time interval the number of muons that decay is proportional to the number of muons present at the start of the interval.

The first part of the experiment took place on a mountain top, at 6300 feet of altitude. One measurement was the number of muons present per hour. In the other the muons were brought to a stop in the detector and the decay times measured. The results showed that about 35% of the muons decayed during each μ-second. This measurement established a relationship between time and the number of muons remaining after a given time interval. So the muon count acts as a clock, and the measure of 'stopped' muons means the clock is at rest.

The second part of the experiment took place at sea level. Measurement of the muon count and use of the muon- time relation showed that about 0.7 μ-seconds had elapsed on the muon clock as it moved from 6300 feet altitude to sea level.

The Lorentz transforms can be applied to this data. Let x,ct refer to the frame at rest in the earth, and x′, ct′ refer to the moving frame in which the muon clock is at rest. Put the common origins at 6300 feet altitude and direct the positive x, x′ axes downward. Using the inverse transforms, p. 25, and setting x′ = 0 to indicate the position of the muon clock leads to the relations

$$x = m(v/c)\, ct', \qquad\qquad ct = m\, ct'.$$

Setting $x = 1.92$ km and $ct' = 0.21$ km in the first equation results in $v/c = 0.994$. Entering this value in m and $ct' = 0.21$ in the second equation results in $ct = 1.93$ km, which corresponds to 6.44 μ-seconds. Thus the measured time interval of the clock moving over a measured path length results in a much larger time interval, as viewed from the stationary frame. This result corresponds to Case 2 of the theoretical calculation of 'slow clocks'.

This example illustrates the relativity principle that two observers in steady relative motion, when viewing the same events, each see a different configuration of the events.

The Polygon. The analysis of Ref. 13 continues by identifying two points A and B, where synchronous clocks are located. The coordinates of the clocks are not specified, indicating that the locations are arbitrary. When the clock at A is moved to B, it lags behind B by the same amount as given for the 'slow clock'. The slow clock moves exclusively along the x axis, whereas the points A and B are allowed to be located anywhere. The picture suggested has 'local' frames with their x axes lined up along the line AB.

The clock is then allowed to move along a series of connected lines forming a polygon, and the polygon is closed by locating B and A at the same place. The local frames must be allowed to align with each segment in turn, as illustrated in Fig 16. The clocks are designated C1 and C2 instead of A and B.

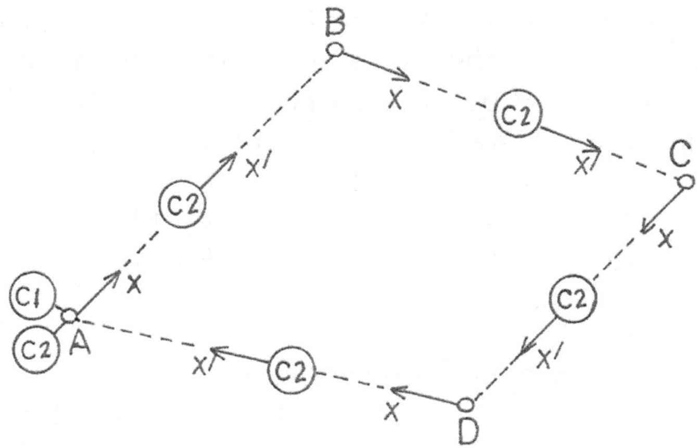

Figure 16 - Path of Moving Clock C2.

The analysis concludes that the clock moving around the closed path is also slow by the same amount as the slow clock when the speed v and the time t taken by the circuit, i.e. the path lengths, are the same.

The conclusion that the time on the moving clock C2 is the same along the segmented path of fig 16 as for movement along a single straight line suggests that the properties of the straight line path should govern the segmented path. In particular, the straight path has no corners such as at B, C, and D therefore the corners should have no effect on the polygon clock.

Since both frames are inertial along the straight path, both should be treated as inertial around the circuit. In which case, the symmetry between the frames is preserved, and C2 can be considered as at rest. And each twin therefore views the other as being younger.

The derivation of the slow clock formula has been described in detail, and experimental support for it has been presented. So it can be said confidently that the theory, as applied here, predicts that, as viewed by the stationary twin, the traveling twin will be younger when he returns home.

Satellites This theory can be applied to the rotating earth and satellites. Some examples are given in Table 3.

Table 3 Some rotating systems

	Earth	ISS	GPS	Geo-stat
Altitude mi.	0	260	12,424	22,286
Speed Mph	1034	17,100	8700	7000
Period hrs	24	1.55	11.8	24
v/c x10e-6	1.54	25.5	13	10.5
Δt Sec x10e-6	0.1	1.8	3.6	4.7

Earth, a clock on the surface, earth rotation speed
ISS, the International Space Station
GPS, Global Positioning System
Geo stat, A satellite that orbits above an earth point
Δt, time lost per revolution

The time lost is calculated by the approximate formula

$$\Delta t = (v /c)^2 t/2,$$ where t is the period in seconds.

Time lost per day, 28 x10e-6, is the most for the Space Station because of the short period of revolution. All of the time lost values are small, compared to the muon times.

But the time lost is still very significant for the GPS satellites because of the precision required for position measurement, and the accumulation of error over time, Refs 21 and 22.. Reference 21 reports that the precision is obtained by adjusting the rate of the orbiting clocks so that their rate, when viewed from the ground, matches the rate of the ground clocks. The demonstrated accuracy of GPS systems suggests the validity of the relativity theory.

Internet searches indicate that a variety of tests have been carried out with orbiting clocks, with results supporting Relativity.

Comparison of orbiting clocks with ground clocks involves gravitational effects of General Relativity, in addition to the effects of Special Relativity. Reference 21 indicates that the time correction for gravitation is much larger than the correction for Special Relativity.

An experiment to test the time loss values of Special Relativity separately would involve two clocks at the same gravitation level. Reference 13 suggests that a clock at rest at the equator would be slow compared to a clock at the earths pole. The analysis of the muons shows that a specific spatial interval is needed to extablish the start and end times of the clocks. A reference point to mark the ends of an interval might be established by a line starting at the center of the earth, extending radially outward to the equator, and then on to a distant fixed star. The passage of a clock across this line is an event both equator and pole observers can use to start their clocks. Subsequent passages after one or more orbits would mark the end of the time interval on each clock. Observation over a long period, such as a year, might improve the accuracy situation.

Coming Home. But the theory also predicts, apparently, that each twin, when moving, is viewed as younger by the stationary twin. This is resolved as follows.

In the section 'Dimensions', page20, it was concluded that the clocks of both frames must start from zero together and they must proceed to mark time at the same rate. Therefore a particular time interval, such as a twins age, marked on the clocks of one frame will match exactly the age marked off on the clocks of the other frame. In this sense the twins are always the same age. The apparent differences when viewed from the other frame are a result of the definition of time and the constant speed property of light. According to Ref. 18, p. 105, time dilation is related to simultaneity and the methods of measurement.

In the derivation of the slow clock formula the moving clock is still moving when it arrives back home. But coming home means stopping, not just driving by. Relativity doesn't say much about how the relative speed v comes about or what happens when it changes. The interval equations (27) can help. Referring to Case 2, p.49, $x1' = x2'$ indicates that the clock of the 'moving' frame is moving. The time interval equation is

$$(ct2' - ct1') = (ct2 - ct1) \sqrt{(1 - v^2 / c^2)}.$$

So the time of the moving clock is less than the time of the stationary clock; the moving twin appears younger. The interval can be something like the twins age or the duration of the whole circuit around the polygon.

The difference in the intervals is determined by the speed v. We can imagine the space traveler arriving home with a gradual descent, like an airliner approaching the airport at gradually decreasing speed. As the speed decreases the difference of the intervals decreases, and when the traveler finally comes to rest the twins are reunited at the same age.

10. The Relativity of Simultaneity

An important new result of Relativity theory is that two events that are simultaneous with respect to one coordinate system can not also be simultaneous with respect to another system in steady relative motion. This principle is significant enough to warrant special detailed description. To high-light the special properties of light postulated by relativity, calculations will be done for both sound and light waves.

A Single Source

The _apparatus_ is shown in figure 17. A train car is moving along its tracks on the ground. The car is open and does not disturb the surrounding air as it moves. There is an observer on the car and one on the ground. The car has clocks (the C in a circle) at the rear, center and front, a source (the S in a circle) at the center, and receivers (the R in a circle) at the front and back ends. There are clocks and receivers also on the ground, as shown. The geometry is arranged so that the source emits a pulse exactly when the position of the center clocks coincide, so that both observers see the pulse originating at their center clock. The arrival of the pulse at the front and back clocks are the events whose times are to be calculated.

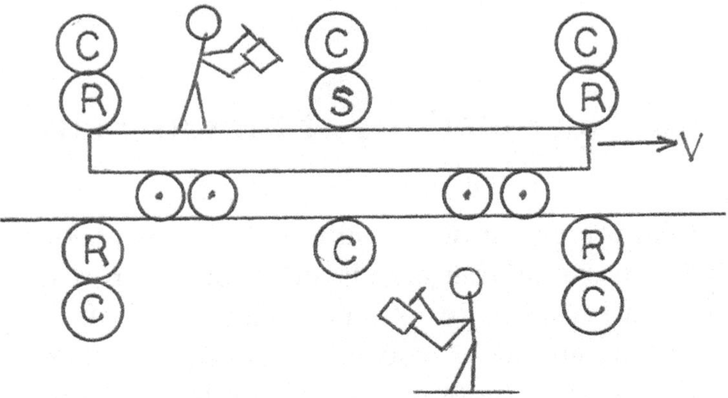

Figure 17.- Apparatus for Simultaneity Study: One source. S = Source, C = Clock, R = Receiver

An _acoustic_ pulse is generated by the source. The speed of this sound pulse is a constant with respect to the air, which is at rest with respect to the ground, and does not depend on the motion of the source. The sound travels at the same speed in both directions, and therefore arrives at the ground end clocks at the same time, i.e. simultaneously. Calculation of the arrival times at the ends of the moving car, with respect to the ground observer, must account for the speeds of both the car and the sound. The result shows that the sound reaches the back end before it reaches the front end because the back and the sound are approaching each other, whereas the front is moving away ahead of the approaching sound. (Think of riding on a train and watching cars on a parallel road. The car moving opposite to the trains motion quickly disappears to the back, while the car moving with the train takes a long time to pass.)

The observer on the car is aware that he is moving through the air because he can feel and measure the wind blowing. He knows that sound speed is constant with respect to the air. He therefore does the same calculation that the ground observer does and finds the sound reaching the back end before it reaches the front, i.e. the two events are not simultaneous.

A _light_ pulse is emitted by the source. There is no light medium, equivalent to the air for the sound, to carry the light wave. Instead the Postulate of Constant Light Speed governs the behavior of the light. The postulate states that the speed of light has the same constant value in all directions and is independent of the (steady) motion of the source or receiver. The ground observer therefore sees the light approaching his end receivers at the same speed, arriving at the same time, i.e. simultaneously. As before with the sound pulse, the ground observer sees* the light and the train car both moving, and finds that, by the same reasoning as with sound, the light reaches the back of the car before it reaches the front. Thus, the ground observer sees the same phenomenon for both the light and sound pulses. The effects are of different magnitudes due to the different speeds of sound and light.

The train car observers situation is now different than it was with the sound pulse. With no medium to anchor the light, the light postulate now says that the light speed is the same for the car as for the ground observer. So the car observer sees the light approaching his end receivers at the same

---

* The term "the observer sees" is shorthand for the more cumbersome "with respect to the coordinate system".

speed and reaching the front and back of the car at the same time, i.e. simultaneously. Thus the two events, the arrival of the light pulse at the front and back of the car, are seen as simultaneous by the car observer, but are seen as not simultaneous by the ground observer.

The car observer can now consider himself to be at rest and the ground observer to be in motion. Using the same reasoning as used by the ground observer, the car observer can calculate the time of arrival of the light pulse at the ground observers end receivers. He finds them to be not simultaneous, whereas the ground observer sees them as simultaneous.

The _conclusion_ is that the simultaneity of two events such as considered here depends on which observer is judging them. These principles have been used in the analysis of the Lorentz transforms. There are other events, such as the flash of two light sources- one at each end of the train car - that require a separate analysis to evaluate their relative simultaneity.

<u>Two Sources</u>

The relativity of simultaneity has also been analyzed with two sources that emit light pulses toward each other from a distance apart. The apparatus for this condition is illustrated in Figure 18.

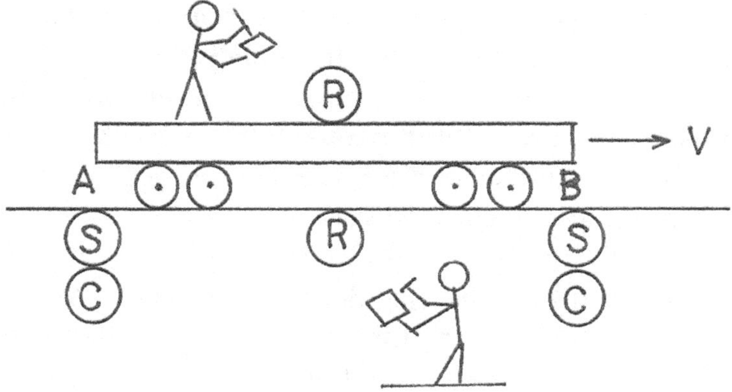

Figure 18.- Setup for simultaneity study: two sources.

S = source, C = clock, R = receiver

The sources at A and B each release a pulse of light toward the receivers R, located at the mid-point of the distance from A to B. Consider first the analysis from the viewpoint of the 'ground' observer, who considers himself to be at rest. The sources are adjusted, perhaps using the clocks, so that the pulses both reach the ground receiver R at the same time, thus they are simultaneous with respect to the ground receiver. The 'ground' observer then asks whether the pulses are also simultaneous with respect to the moving receiver. In the view of the ground observer they are not, for the reason that the moving observer is advancing toward the pulse coming from B, but is moving ahead of the pulse coming from A. This conclusion results from the condition that the 'ground' observer 'sees' both the light pulses and the rail car to be in motion.* Thus, the pulse from B reaches the

---

* This principle was used in analyzing the Lorentz transforms.

'moving' receiver before the pulse from A can catch up, i.e. not simultaneously. It is concluded from this that, in the view of the 'ground' observer, two light pulses that are simultaneous with respect to his receiver are not also simultaneous with respect to the 'moving' rail car receiver.

Relativity also asserts that the rail car observer is equally entitled to consider himself to be at rest. If the sources are then adjusted to be simultaneous with respect to the 'moving' receiver, it follows from the above discussion that, in the view of the rail car observer, they will not be simultaneous in respect to the 'ground' receiver.

Graphing

The following is an alternate explanation that uses graphing of the rays to help visualize the process.

One consequence of Special Relativity is the appearance of a new concept of time. Previously time was said to flow of its own accord and to have the same value for all observers, whatever their state of motion. Relativity says that each observer has their own time in relation to other observers. The train example is intended to give a physical feel for the origin of this time.

Consider an observer standing near the tracks and a train passing at a steady speed with an observer on board at the mid point of the length of the train. When the observers coincide a light flashes at the midpoint and send rays toward the front and rear of the train.

It may help to graph as we go along. Place x horizontal with values from -1 to +2 and ct vertical with values from 0 to +2. Points are noted by (x, ct). The train extends from x = -1 to x = +1.

The train observer considers himself to be at rest and sees each ray moving at ct. The front ray moves at x = ct from (0, 0) to (1, 1), arriving at the front x = 1 at ct = 1. (Label this line L+) The rear ray moves at x = - ct from (0, 0) to (-1, 1), arriving at the rear x = -1 at ct = 1. (Label this line L-) Thus each ray arrives at its end at ct = 1, i.e. the two events are simultaneous.

But when viewed by the track observer both the light and the train are moving. The track observer sees the light rays moving at ct, the same as the train observer, so L+ and L- apply. The rear of the train is moving along the line x = -1 +.5ct, for v =.5c, from (-1, 0) to (-.5, 1). (Label this line B) The rear ray reaches the back of the train where B and L- intersect, at (-2/3, 2/3) The front of the train moves along the line x = 1 +.5ct, and intersects L+ at (2, 2). Thus the two events are *not* simultaneous as viewed by the track observer, the rear ray arriving at ct = 2/3 and the front ray arriving at ct = 2.

Many writers are satisfied that this intuitive analysis proves the Relativity of Simultaneity, perhaps noting that it does not account for 'time dilation' or 'length contraction'. These effects are taken into account by using the Lorentz Transforms for the analysis. The procedure is to transform the points (-1, 1), (0, 0), and (1, 1) using the form with the + sign. The result is the 'view from the stationary observer'. The transformed points are (-.577,.577), (0, 0), and (1.73, 1.73). These are the quantitative results for this example. The track-stationary observer still sees the events as non-simultaneous.

It is useful to compare the behavior of non-relativistic waves. Let a wind be blowing at the same speed as the train is moving. Sound waves are generated at the midpoint going forward and aft instead of light. The train is then at rest in the air and the sound waves follow symmetrical paths such as L+ and L-. The sound waves arrive at the ends simultaneously as before. But the track observer does not see symmetrical waves as before. The forward sound wave speed is augmented by the speed of the wind, and the rear wave speed is decreased by the wind speed. Using the same methods as above shows that the track observer also sees the sound waves arriving at the ends of the train simultaneously.

11. The Space Time Interval

Consider a graph of ct vs x, with x 'horizontal'. Locate both x,ct and x′,ct′ on the same graph. Draw the line x = ct, indicating the path of a light ray. Choose a grid of values x, ct and calculate the values x′, ct′ corresponding to each x, ct.

For example, x,ct =.1,.9 transforms to x′,ct′ =.01,.89 (v/c =.1). An arrow from x,ct to x′,ct′ points strongly left and a little down. All points starting with x,ct above the light line x = ct transform similarly, downward but more leftward. No such point transforms to a point x′,ct′ below the light line.

For another example, the point x,ct =.9,.1 -→ x′,ct′ =.89,.01. Points below the light line transform strongly downward. None transform to above the line.

The distance a point transforms depends on the value of v/c. Starting with x,ct = 4,6 the transformed values for various values of v/c are shown in Table 4

Table 4, Transform of x,ct = 4,6

v/c	m	x′	ct′
.1	1.005	3.42	5.63
.2	1.02	2.86	5.3
.3	1.05	2.31	5.04
.4	1.09	1.74	4.8
.5	1.155	1.155	4.62
.6	1.25	.5	4.5
2/3	1.342	0	4.47

A graph of these numbers shows a curve starting with zero slope at the value ct′ = 4.47 and increasing slope as x′ increases. The equation of this line is obtained as follows. Start with the inverse Lorentz transform equations, p. 25. Set x′ = 0 and ct′ = ct_0, to get

$$x = mvct_0/c \text{ and } ct = m\, ct_0.$$

These equations are parametric with v/c as the parameter. Re-arranging algebraically to eliminate v/c from appearing explicitly leads to the equation

$$ct^2 - x^2 = ct_0^2. \tag{32}$$

Entering the values x, ct = 4, 6 leads to the value ct_0 = 4.47. The resulting equation matches the data in Table 4.

Equation (32) is referred to as an 'interval', or the 'invariant space time interval', Ref 4. The values x, and ct are coordinates, i.e. distances from the origin of a particular Cartesian frame, analogous to the legs of a triangle. So ct_0 might be thought of like a 'hypotenuse', but the negative sign indicates that its properties will be much different. For example, ct_0 reduces to zero for every point on the light line $x = ct$, whereas a hypotenuse reduces to zero only at one specific point.

A similar equation can be obtained for points below the light line. Start with the Lorentz transform equations, p. 25. Set $x' = 1$ and $ct' = 0$ to get the relations $x = m$ and $ct = vm/c$. Again, doing the algebra to eliminate the parameter v results in the equation

$$x^2 - ct^2 = 1 \qquad\qquad (33)$$

This curve originates at the point $x, ct = 1, 0$. It rises steeply then slopes to the right, toward positive x, to become asymptotic to the light line $x = ct$. It has been called a 'calibration' curve because it shows that the point $x' = 1$ is located at a value greater than x, indicating that the scale is different on the two axes.

Direct substitution of the Lorentz transforms into equation (32) leads to the equation:

$$ct^2 - x^2 = ct'^2 - x'^2. \qquad\qquad (34)$$

Equation (34) is presented as a means of computation. If three of the quantities are known from measurement then the fourth can be found from the equation. Consider Table 4, $x, ct = 1.74, 4.8$. If $x' = .5$, what is ct'? Entering these values in eqn.(34) results in $ct = 4.5$, in agreement with the value in Table 4, which was calculated by a different means.

Table 4 shows that the two points considered, $x, ct = 1.74, 4.8$ and $x, ct = .5, .45$, are each associated with a different value of v/c. A single value of v/c that relates these points is obtained by substituting both values in the Lorentz equations. The result is v/c =.26. So a specific pair of points can be related by two different values of v/c,.4 and.6 in Table 4, or by a single value,.26 by the transforms.

Returning to the muons, the experiment showed that $x' = 0$, $ct' =.21$ (0.7 μsec) and $x = 1.92$m. Applying eqn (33) leads to $ct = 1.93$ (6.4 μsec). It can be shown that the value of v/c that reduces the value of x of a given pair x, ct to zero is given by x/ct. Note in Table 4 that the data start with the point 4, 6 and x is zero at v/c =2/3. For the muons x, ct = 1.92, 1.93. Applying the rule the v/c of the muons is 1.92/1.93, or v/c =.995. These results agree with the results obtained earlier using the Lorentz equations.

Reference 4 identifies the interval, equations (32) and (34), as the 'central theme' of that book. Reference 23 derives a similar equation, called the 'metric equation',

$$s^2 = ct^2 - x^2 - y^2 - z^2$$

(in current notation) that is described as the key to understanding the special features of relativity. These books are a subject for another day.

Afterword

The purpose of this work was to clarify my understanding of the Theory of Special Relativity as presented in the paper 'On the Electrodynamics of Moving Bodies by A. Einstein in 1905. I think I have succeeded. I hope that what is written here will help readers who are also new to Special Relativity.

Some readers may say of this book that 'oh, everybody already knows all that'. Some people must already know Special Relativity, because they are building real-world hardware, such as GPS, that depends on Relativity and that hardware works. But the view presented here was very difficult to arrive at from the current literature.

JSM
Nov 2013

12. References

1. The Patent Clerks Legacy. Gary Stix. Scientific American. Volume 291, No. 3, September 2004

2. Einstein [in a nutshell]. Michio Kaku. Discover Magazine, September 2004

3. The Universe in a Nutshell. Stephen Hawking. Bantam Books, November 2001

4. Spacetime Physics: Introduction to special relativity, 2nd edition. E. F. Taylor and J. A. Wheeler. W.H. Freeman & Co. 1992 (ninth printing 2003)

5. Six not-so-easy Pieces. R.P. Feynman. Perseus Books 1997. Page 50.

6. The Mathematical Principles of Natural Philosophy. Isaac Newton. Citadel Press 1964, page28.

7. The Interference of Light. Thomas Young Reprinted in 'Great Experiments in Physics', H. Holt & Co. 1959

8. Michelson and the Speed of Light. B. Jaffe.Anchor Books,1960, page79

9. A Brief History of Time. Stephen Hawking. Bantam Books1998, page 19.

10. The Electromagnetic Field. J. C. Maxwell, Reprinted in 'Great Experiments in Physics, page 283

11. H. A. Lorentz. Several articles in 'The Principle of Relativity' Dover Pubs. 1923

12. Henri Poincaré, Article from internet, Wikipedia

13. On the Electrodynamics of Moving Bodies. A. Einstein, in 'The Principle of Relativity', Dover Publications, p.37.

14. Electromagnetic Phenomena in a System Moving with any Velocity less than that of Light. H. A. Lorentz. In 'The Principle of Relativity', Dover Publications. p. 11

15. The Electromagnetic Field. A. Shadowitz. Dover Publications, 1975. p.57.

16. Fundamental Ideas and Problems of the Theory of Relativity. A. Einstein. Lecture delivered to the Nordic Assembly of Naturalists at Gothenberg. July 11, 1923.

17. The Meaning of Relativity. A. Einstein. Princeton Univ. Press. 1953. p. 38

18. Special Relativity. A..P. French, W.W. Norton Co. p 114, also p 96 'Moving objects appear…contracted', also p149-152 the moving board.

19. Relativity: The Special and the General Theory. A. Einsteein, Bonanza Books, 1961. P 35-36

20. Length measurement of a moving rod by a single observer without assumptions concerning

its magnitude. Bernard Rothenstein and Ioan Camian. Obtained from interner at arXiv: thysics/0507016, 2005

21. Real World Relativity: Thhe GPS Navigation System. Richard W. Pogge, www.astronomy.ohio-state.edu/...gps.ht...

22. Relativity and The Global Positioning System. Neil Ashby, Physics Today, vol. 55, May, 2002

23. A Travelers Guide to Spacetime. Thomas A. Moore. McGraw-Hill, 1995

Made in the USA
Monee, IL
07 July 2026

56552684R00049